α Alpha

Single-Digit Addition and Subtraction

Tests

Math·U·See

1-888-854-MATH (6284) - www.MathUSee.com
Sales@MathUSee.com

Alpha Tests: Single-Digit Addition and Subtraction
©2012 Math-U-See, Inc.
Published and distributed by Demme Learning

www.MathUSee.com

1-888-854-6284 or +1 717-283-1448 | www.demmelearning.com
Lancaster, Pennsylvania USA

ISBN 978-1-60826-072-0
Revision Code 0616

Printed in the United States of America by Bindery Associates LLC

For information regarding CPSIA on this printed material call: 1-888-854-6284
and provide reference #0616-07272016

Turn the paper sideways. Color the right number of blocks and say the answer.

1.

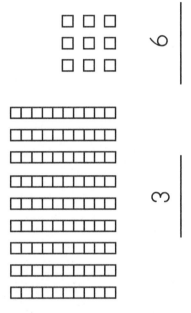

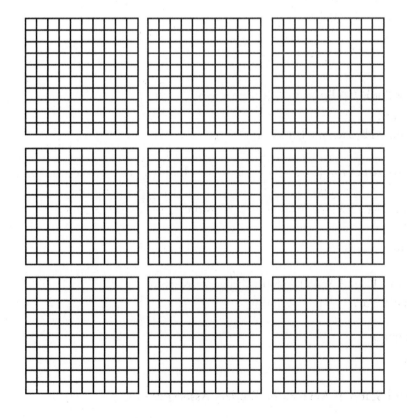

Count and write the number. Then say it.

2.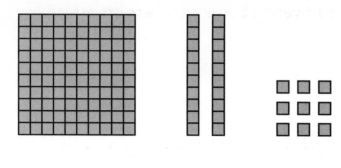

_____   _____   _____

Build and say the number.

3. 141   4. 390

5. What is the greatest number of units that can live in the units house?

_____

6. What is the greatest number of tens that can live in the tens house?

_____

LESSON TEST

Count and write from zero to twenty. Some numbers are already
written for you.

1.

  — — — — — — — — —

  1 0 — — — — — — — —

  —

2.

  — — — — — — — 7 — —

  — — — — — — — — —

  —

Count and write the number. Then say it.

3.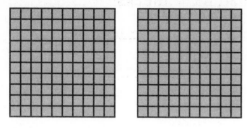

_____   _____   _____

Build and say the number.

4. 157

5. 38

3

Match and say. Color the squares to match the blocks.

1.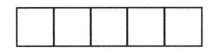

2.

3.

4.

Write or say the answer.

5. What color is the eight block? _____

Count and write the number. Then say it.

6.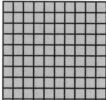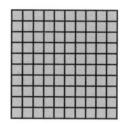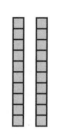

   _____   _____ _____

Count and write from zero to twenty. Some numbers are already written for you.

7.

   __ __ __ __ __ __ __ __ __ __

   __ __ __ __ __ __ __ __ _18_ __

   __

Solve.

1.
$$\begin{array}{r} 0 \\ +\ 1 \\ \hline \end{array}$$

2.
$$\begin{array}{r} 8 \\ +\ 0 \\ \hline \end{array}$$

3.
$$\begin{array}{r} 0 \\ +\ 6 \\ \hline \end{array}$$

4.
$$\begin{array}{r} 5 \\ +\ 0 \\ \hline \end{array}$$

5.
$$\begin{array}{r} 4 \\ +\ 0 \\ \hline \end{array}$$

6.
$$\begin{array}{r} 0 \\ +\ 3 \\ \hline \end{array}$$

7. $0 + 0 =$ _____

8. $0 + 7 =$ _____

9. $2 + 0 =$ _____

10. $1 + 0 =$ _____

11. $3 + 0 =$ _____

12. $0 + 5 =$ _____

Build and say the number.

13. 350

14. 102

15. What color is the nine block? _____

16. David ate 6 pieces of candy. His dad said, "No more!" How much candy did David eat?

_____

5

Solve.

1.
$$\begin{array}{r} 5 \\ + \ 1 \\ \hline \end{array}$$

2.
$$\begin{array}{r} 4 \\ + \ 1 \\ \hline \end{array}$$

3.
$$\begin{array}{r} 1 \\ + \ 8 \\ \hline \end{array}$$

4.
$$\begin{array}{r} 2 \\ + \ 1 \\ \hline \end{array}$$

5.
$$\begin{array}{r} 7 \\ + \ 1 \\ \hline \end{array}$$

6.
$$\begin{array}{r} 1 \\ + \ 3 \\ \hline \end{array}$$

7. $1 + 1 =$ _____

8. $1 + 9 =$ _____

9. $6 + 1 =$ _____

10. $0 + 5 =$ _____

11. $8 + 0 =$ _____

12. $0 + 2 =$ _____

Count and write the number. Then say it.

13.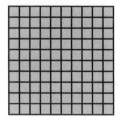

_____   _____   _____

Build and say the number.

14. 214

15. There are five girls and one boy in Jen's family. How many children are there in all?

_____

6

Count and write from 0 to 100.

1.

0 __ __ __ __ __ __ __ __ __

__ __ __ __ __ __ __ __ __ 19

__ __ 22 __ __ __ __ __ __

__ __ __ __ __ 35 __ __ __ __

40 __ __ __ __ __ __ __ __ __

__ __ __ __ __ __ __ __ __ __

__ __ __ __ __ __ __ 67 __ __

__ 71 __ __ __ __ __ __ __ __

__ __ __ __ __ __ __ __ __ __

__ __ __ 93 __ __ __ __ __ __

__

Skip count by 10 and write the numbers.

2. ___, ___, 30, ___, ___, ___, 70, ___, ___, ___

Solve.

3.   1
   +   6

4.   5
   +   1

5.   9
   +   0

6.   1
   +   8

LESSON TEST

Solve.

1.
```
     1
 +   2
-------
```

2.
```
     2
 +   4
-------
```

3.
```
    20
 +  20
-------
```

4.
```
   100
 + 100
-------
```

5.
```
     6
 +   2
-------
```

6.
```
     5
 +   2
-------
```

7.
```
     2
 +   7
-------
```

8.
```
     4
 +   2
-------
```

9. $2 + 3 = $ _____   10. $2 + 6 = $ _____

11. $0 + 2 = $ _____   12. $1 + 7 = $ _____

13. $3 + 0 = $ _____   14. $6 + 1 = $ _____

15. $9 + 1 = $ _____   16. $0 + 4 = $ _____

17. $1 + 2 = $ _____

18. Mia bought 3 red pencils and 2 green pencils. How many pencils did she buy?

_____

Solve for the unknown. Use the blocks if needed.

1. _____ + 1 = 1

2. _____ + 0 = 4

3. _____ + 2 = 2

4. _____ + 0 = 8

5. _____ + 2 = 5

6. _____ + 1 = 7

7. _____ + 2 = 3

8. _____ + 1 = 6

9. _____ + 0 = 0

10. _____ + 1 = 9

11. _____ + 0 = 1

12. _____ + 2 = 4

Solve.

13.
$$6$$
$$+ \ 2$$

14.
$$7$$
$$+ \ 1$$

15.
$$30$$
$$+ \ 20$$

16.
$$9$$
$$+ \ 0$$

17. Jordan has 3 model cars. He wants to have 5 cars. How many more cars does he want?

_____

18. There are 9 players on a baseball team. We have 8 players. How many more players do we need to make one team?

_____

LESSON TEST

Solve.

1.
$$\begin{array}{r} 0 \\ + \ 9 \\ \hline \end{array}$$

2.
$$\begin{array}{r} 9 \\ + \ 7 \\ \hline \end{array}$$

3.
$$\begin{array}{r} 6 \\ + \ 9 \\ \hline \end{array}$$

4.
$$\begin{array}{r} 9 \\ + \ 9 \\ \hline \end{array}$$

5.
$$\begin{array}{r} 9 \\ + \ 2 \\ \hline \end{array}$$

6.
$$\begin{array}{r} 3 \\ + \ 9 \\ \hline \end{array}$$

7. 9 + 1 = _____

8. 8 + 9 = _____

9. 9 + 7 = _____

10. 4 + 9 = _____

11. 6 + 1 = _____

12. 7 + 2 = _____

Solve for the unknown. Use the blocks if needed.

13. _____ + 9 = 11

14. _____ + 2 = 6

15. _____ + 1 = 4

Skip count by 10 and write the numbers.

16. 10, ____ , ____ , ____ , 50, ____ , ____ , ____ , ____ , ____

17. Jed read 9 books last week and 8 books this week. How many books did he read in all?

_____

18. Alexis has six dollars. She needs eight dollars to buy a game. How many more dollars does she need?

_____

LESSON TEST

Solve.

1.
$$
\begin{array}{r}
8 \\
+8 \\
\hline
\end{array}
$$

2.
$$
\begin{array}{r}
8 \\
+5 \\
\hline
\end{array}
$$

3.
$$
\begin{array}{r}
0 \\
+8 \\
\hline
\end{array}
$$

4.
$$
\begin{array}{r}
2 \\
+8 \\
\hline
\end{array}
$$

5.
$$
\begin{array}{r}
8 \\
+6 \\
\hline
\end{array}
$$

6.
$$
\begin{array}{r}
8 \\
+3 \\
\hline
\end{array}
$$

7.
$$
\begin{array}{r}
80 \\
+10 \\
\hline
\end{array}
$$

8.
$$
\begin{array}{r}
7 \\
+8 \\
\hline
\end{array}
$$

9. $1 + 7 =$ _____   10. $8 + 9 =$ _____

11. $2 + 5 =$ _____   12. $9 + 7 =$ _____

13. $1 + 3 =$ _____   14. $9 + 6 =$ _____

15. $4 + 9 =$ _____   16. $7 + 2 =$ _____

Solve for the unknown. Use the blocks if needed.

17. $X + 7 = 15$ 18. $X + 9 = 18$

19. Kayla rode her bike 8 miles and walked 5 miles. How far did she travel?

_____

20. Emily needs 17 beads to make a necklace. She has 9 beads. How many more beads does she need?

_____

Solve.

1.
$$\begin{array}{r} 7 \\ +\ 9 \\ \hline \end{array}$$

2.
$$\begin{array}{r} 2 \\ +\ 2 \\ \hline \end{array}$$

3.
$$\begin{array}{r} 4 \\ +\ 9 \\ \hline \end{array}$$

4.
$$\begin{array}{r} 2 \\ +\ 5 \\ \hline \end{array}$$

5.
$$\begin{array}{r} 6 \\ +\ 2 \\ \hline \end{array}$$

6.
$$\begin{array}{r} 8 \\ +\ 7 \\ \hline \end{array}$$

7.
$$\begin{array}{r} 20 \\ +\ 10 \\ \hline \end{array}$$

8.
$$\begin{array}{r} 8 \\ +\ 0 \\ \hline \end{array}$$

9.
$$2$$
$$+ \quad 9$$

10.
$$9$$
$$+ \quad 9$$

11.
$$70$$
$$+ \quad 20$$

12.
$$100$$
$$+ \quad 800$$

13. $5 + 8 =$ _____

14. $8 + 4 =$ _____

15. $8 + 8 =$ _____

16. $2 + 4 =$ _____

17. $8 + 6 =$ _____

18. $6 + 9 =$ _____

19. $2 + 3 =$ _____

20. $9 + 5 =$ _____

21. $2 + 8 =$ _____

22. $9 + 9 =$ _____

23. $3 + 9 =$ _____

24. $9 + 8 =$ _____

Solve for the unknown. Use the blocks if needed.

25. $X + 7 = 15$ 26. $X + 9 = 18$

27. $X + 8 = 17$

Count and write the number. Then say it.

28.
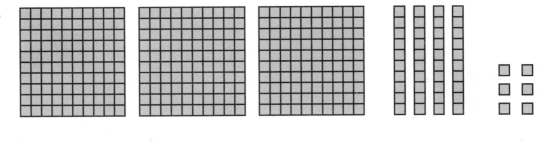

_____   _____   _____

Count and write from zero to twenty. Some numbers are already written for you.

29.

0 — — — — — — — — —

— — — — — — — — —

20

Skip count by 10 and write the numbers.

30. 10, _____ , _____ , _____ , 50, _____ , _____ , _____ , _____ , _____

31. I saw 8 robins and 3 blue jays in my yard. How many birds did I see?

   _____

32. Michael found five pennies under the bed and one penny in the closet. How many pennies did Michael find?

   _____

Use the drawing to answer the questions.

1. How many circles are striped? _____

2. How many triangles are plain? _____

3. How many circles are plain? _____

4. How many triangles are striped? _____

Solve.

5.
$$9$$
$$+\ \ 8$$

6.
$$8$$
$$+\ \ 5$$

7.
$$2$$
$$+\ \ 7$$

8.
$$40$$
$$+\ 20$$

9. $3 + 0 =$ _____

10. $4 + 2 =$ _____

11. $9 + 9 =$ _____

12. $6 + 2 =$ _____

Solve for the unknown. Use the blocks if needed.

13. $A + 8 = 14$

14. $2 + X = 7$

Skip count by two and write the numbers.

15. 2, ____ , ____ , ____ , 10, ____ , ____ , ____ , ____ , ____

12

Solve.

1.
$$\begin{array}{r} 8 \\ +\ 8 \\ \hline \end{array}$$

2.
$$\begin{array}{r} 5 \\ +\ 5 \\ \hline \end{array}$$

3.
$$\begin{array}{r} 30 \\ +\ 30 \\ \hline \end{array}$$

4.
$$\begin{array}{r} 4 \\ +\ 4 \\ \hline \end{array}$$

5.
$$\begin{array}{r} 6 \\ +\ 6 \\ \hline \end{array}$$

6.
$$\begin{array}{r} 9 \\ +\ 9 \\ \hline \end{array}$$

7.
$$\begin{array}{r} 7 \\ +\ 7 \\ \hline \end{array}$$

8.
$$\begin{array}{r} 2 \\ +\ 2 \\ \hline \end{array}$$

9. $5 + 8 =$ _____

10. $9 + 4 =$ _____

11. $3 + 2 =$ _____

12. $6 + 9 =$ _____

Solve for the unknown. Use the blocks if needed.

13. $B + 9 = 18$

14. $Y + 8 = 15$

15. $Q + 2 = 11$

Count the number of each shape in the box.

16. There are _____ circles.

17. There are _____ triangles.

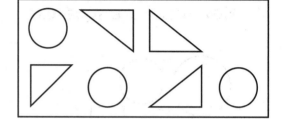

18. John went fishing with his dad. He caught 4 fish, and his dad caught 4 fish. How many fish were caught in all?

_____

19. Ali solved 6 problems on her math worksheet. There are 15 problems in all. How many does Ali have left to solve?

_____

20. Chuck has 7 chores to do for his mom and 7 chores to do for his dad. How many chores does he have to do in all?

_____

Use the drawings to answer the questions.

1. How many are rectangles? _____

2. How many are squares? _____

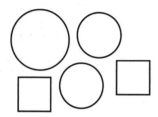

3. How many are triangles? _____

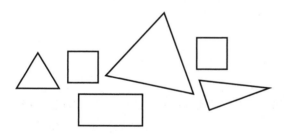

4. How many are circles? _____

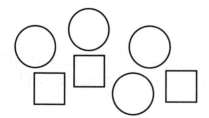

Solve.

5.
$$
\begin{array}{r}
6 \\
+\ 6 \\
\hline
\end{array}
$$

6.
$$
\begin{array}{r}
3 \\
+\ 3 \\
\hline
\end{array}
$$

7.
$$
\begin{array}{r}
7 \\
+\ 7 \\
\hline
\end{array}
$$

8.
$$
\begin{array}{r}
20 \\
+\ 40 \\
\hline
\end{array}
$$

9. $9 + 0 =$ _____

10. $7 + 1 =$ _____

11. $8 + 9 =$ _____

12. $5 + 8 =$ _____

Solve for the unknown. Use the blocks if needed.

13. $A + 9 = 15$

14. $4 + X = 8$

15. Ashley drew a square on her paper.
How many sides did it have? _____

Skip count by five and write the numbers.

16. ____ , ____ , ____ , 20, ____ , ____ , ____ , ____ , 45, ____

ALPHA

Solve.

1.
$$\begin{array}{r} 50 \\ +\ 40 \\ \hline \end{array}$$

2.
$$\begin{array}{r} 8 \\ +\ 7 \\ \hline \end{array}$$

3.
$$\begin{array}{r} 1 \\ +\ 2 \\ \hline \end{array}$$

4.
$$\begin{array}{r} 6 \\ +\ 7 \\ \hline \end{array}$$

5.
$$\begin{array}{r} 2 \\ +\ 3 \\ \hline \end{array}$$

6.
$$\begin{array}{r} 5 \\ +\ 6 \\ \hline \end{array}$$

7.
$$\begin{array}{r} 8 \\ +\ 9 \\ \hline \end{array}$$

8.
$$\begin{array}{r} 400 \\ +\ 300 \\ \hline \end{array}$$

9. $7 + 7 = $ _____

10. $8 + 3 = $ _____

11. $6 + 6 = $ _____

12. $9 + 5 = $ _____

Write the answers in words.

zero	six
one	seven
two	eight
three	nine
four	ten
five	

13. Three plus two equals _____ .

14. Two plus two equals _____ .

Circle the name of the shape.

15.
square

circle

triangle

16.
square	rectangle
circle	triangle

17. Timothy has 5 toes on his left foot and 5 toes on his right foot. How many toes does Timothy have?

_____

18. Caleb has five sisters and one brother. How many brothers and sisters does he have?

_____

Counting Caleb, how many children are in the family?

_____

Solve.

1.
$$\begin{array}{r} 6 \\ +\ 4 \\ \hline \end{array}$$

2.
$$\begin{array}{r} 9 \\ +\ 7 \\ \hline \end{array}$$

3.
$$\begin{array}{r} 10 \\ +\ 20 \\ \hline \end{array}$$

4.
$$\begin{array}{r} 5 \\ +\ 5 \\ \hline \end{array}$$

5.
$$\begin{array}{r} 9 \\ +\ 1 \\ \hline \end{array}$$

6.
$$\begin{array}{r} 6 \\ +\ 8 \\ \hline \end{array}$$

7.
$$\begin{array}{r} 8 \\ +\ 2 \\ \hline \end{array}$$

8.
$$\begin{array}{r} 8 \\ +\ 8 \\ \hline \end{array}$$

9. $7 + 3 =$ _____

10. $3 + 4 =$ _____

11. $5 + 2 =$ _____

12. $6 + 4 =$ _____

13. $5 + 6 =$ _____

14. $7 + 8 =$ _____

15. $9 + 5 =$ _____

16. $7 + 6 =$ _____

Solve for the unknown. Use the blocks if needed.

17. $B + 6 = 10$

18. $Y + 5 = 10$

19. $Q + 3 = 10$

20. Lisa's mother said she had to pick up 10 toys. Lisa has already picked up 8 toys. How many more toys does she have to pick up?

_____

Solve.

1.
$$\begin{array}{r} 2 \\ +\ 7 \\ \hline \end{array}$$

2.
$$\begin{array}{r} 5 \\ +\ 5 \\ \hline \end{array}$$

3.
$$\begin{array}{r} 10 \\ +\ 80 \\ \hline \end{array}$$

4.
$$\begin{array}{r} 6 \\ +\ 3 \\ \hline \end{array}$$

5.
$$\begin{array}{r} 7 \\ +\ 8 \\ \hline \end{array}$$

6.
$$\begin{array}{r} 5 \\ +\ 4 \\ \hline \end{array}$$

7.
$$\begin{array}{r} 6 \\ +\ 6 \\ \hline \end{array}$$

8.
$$\begin{array}{r} 5 \\ +\ 9 \\ \hline \end{array}$$

9. 8 + 1 = _____

10. 4 + 4 = _____

11. 8 + 4 = _____

Solve for the unknown. Use the blocks if needed.

12. 5 + A = 9   13. 6 + X = 9

14. 7 + G = 9

Match each shape with the best name.

15.   triangle

16.   square

17.    rectangle

18.   circle

19. Sara traveled to a faraway star with 9 planets. She has visited 4 of the planets. How many are left to visit?

_____

20. Dan lost 60 pennies yesterday and 30 pennies today. How many pennies has he lost?

_____

LESSON TEST

Solve.

1.
$$\begin{array}{r} 30 \\ + 50 \\ \hline \end{array}$$

2.
$$\begin{array}{r} 4 \\ + 7 \\ \hline \end{array}$$

3.
$$\begin{array}{r} 7 \\ + 5 \\ \hline \end{array}$$

4.
$$\begin{array}{r} 5 \\ + 4 \\ \hline \end{array}$$

5.
$$\begin{array}{r} 5 \\ + 7 \\ \hline \end{array}$$

6.
$$\begin{array}{r} 7 \\ + 3 \\ \hline \end{array}$$

7.
$$\begin{array}{r} 200 \\ + 200 \\ \hline \end{array}$$

8.
$$\begin{array}{r} 8 \\ + 7 \\ \hline \end{array}$$

9. $5 + 5 =$ _____   10. $9 + 9 =$ _____

11. $6 + 4 =$ _____   12. $3 + 8 =$ _____

13. $6 + 3 =$ _____   14. $8 + 5 =$ _____

Solve for the unknown. Use the blocks if needed.

15. $4 + X = 11$ 16. $7 + B = 12$

17. $3 + R = 8$

18. The guests brought six red balloons and five yellow balloons to Mary's party. How many balloons are at the party?

    _____

19. I noticed 4 bluebirds and 3 redbirds at my bird feeder. How many birds are at the feeder?

    _____

Skip count by five and write the numbers.

20. 5, _____ , _____ , _____ , _____ , _____ , _____ , _____ , _____ , _____

Solve.

1.
$$\begin{array}{r} 9 \\ +\ 2 \\ \hline \end{array}$$

2.
$$\begin{array}{r} 3 \\ +\ 3 \\ \hline \end{array}$$

3.
$$\begin{array}{r} 7 \\ +\ 6 \\ \hline \end{array}$$

4.
$$\begin{array}{r} 3 \\ +\ 5 \\ \hline \end{array}$$

5.
$$\begin{array}{r} 5 \\ +\ 1 \\ \hline \end{array}$$

6.
$$\begin{array}{r} 2 \\ +\ 7 \\ \hline \end{array}$$

7.
$$\begin{array}{r} 40 \\ +\ 10 \\ \hline \end{array}$$

8.
$$\begin{array}{r} 2 \\ +\ 4 \\ \hline \end{array}$$

9.
$$\begin{array}{r} 2 \\ +\ 2 \\ \hline \end{array}$$

10.
$$\begin{array}{r} 2 \\ +\ 8 \\ \hline \end{array}$$

Solve.

11. $\begin{array}{r} 50 \\ +\ 20 \\ \hline \end{array}$ 12. $\begin{array}{r} 400 \\ +\ 500 \\ \hline \end{array}$

13. $\begin{array}{r} 8 \\ +\ 6 \\ \hline \end{array}$ 14. $\begin{array}{r} 9 \\ +\ 7 \\ \hline \end{array}$

15. $\begin{array}{r} 40 \\ +\ 30 \\ \hline \end{array}$ 16. $\begin{array}{r} 300 \\ +\ 100 \\ \hline \end{array}$

17. 8 + 3 = _____   18. 6 + 9 = _____

19. 6 + 6 = _____   20. 8 + 4 = _____

21. 0 + 6 = _____   22. 7 + 7 = _____

23. 5 + 6 = _____   24. 8 + 5 = _____

25. $\begin{array}{r} 7 \\ +\ 3 \\ \hline \end{array}$ 26. $\begin{array}{r} 9 \\ +\ 9 \\ \hline \end{array}$

Solve.

27.
```
    9
+   4
─────
```

28.
```
    7
+   8
─────
```

29.
```
    4
+   4
─────
```

30.
```
    2
+   3
─────
```

31.
```
   20
+  30
─────
```

32.
```
    7
+   5
─────
```

33.
```
    0
+   9
─────
```

34.
```
    5
+   5
─────
```

35.
```
   10
+  10
─────
```

36.
```
  600
+ 300
─────
```

37.
$$\begin{array}{r} 8 \\ + \quad 0 \\ \hline \end{array}$$

38.
$$\begin{array}{r} 3 \\ + \quad 9 \\ \hline \end{array}$$

39.
$$\begin{array}{r} 20 \\ + \quad 60 \\ \hline \end{array}$$

40.
$$\begin{array}{r} 300 \\ + \quad 300 \\ \hline \end{array}$$

41. $8 + 8 =$ _____

42. $4 + 7 =$ _____

43. $9 + 8 =$ _____

44. $4 + 6 =$ _____

Circle the name of the shape.

45.

| rectangle |
| circle |
| triangle |

46.

| square | rectangle |
| circle | triangle |

47.

| square |
| circle |
| triangle |

48.

| square | rectangle |
| circle | triangle |

18

Name each bar. Tell if it means have or owe. Color the bars if you wish.

1. ⬜⬜

   _____

2. ▦▦▦▦▦▦▦▦▦▦

   _____

3. ⬜⬜⬜

   _____

4. ⬜⬜⬜⬜

   _____

5. ⬜⬜⬜⬜⬜

   _____

6. ▦▦▦▦▦▦▦▦

   _____

Solve.

7. $\begin{array}{r} 4 \\ + \quad 7 \\ \hline \end{array}$

8. $\begin{array}{r} 6 \\ + \quad 5 \\ \hline \end{array}$

9. $\begin{array}{r} 5 \\ + \quad 4 \\ \hline \end{array}$

10. $\begin{array}{r} 600 \\ + 300 \\ \hline \end{array}$

Solve for the unknown. Use the blocks if needed.

11. $X + 0 = 7$

12. $A + 1 = 3$

13. $Y + 0 = 6$

14. $B + 1 = 9$

15. What number plus three is the same as eight?

_____

16. What number plus six is the same as ten?

_____

LESSON TEST

19

Subtract and write your answer. Use addition to check the answer.

1.
$$\begin{array}{r} 4 \\ -\ 1 \\ \hline \end{array}$$

2.
$$\begin{array}{r} 6 \\ -\ 5 \\ \hline \end{array}$$

3.
$$\begin{array}{r} 8 \\ -\ 8 \\ \hline \end{array}$$

4.
$$\begin{array}{r} 1 \\ -\ 0 \\ \hline \end{array}$$

5.
$$\begin{array}{r} 2 \\ -\ 1 \\ \hline \end{array}$$

6.
$$\begin{array}{r} 4 \\ -\ 3 \\ \hline \end{array}$$

7.
$$\begin{array}{r} 7 \\ -\ 0 \\ \hline \end{array}$$

8.
$$\begin{array}{r} 9 \\ -\ 1 \\ \hline \end{array}$$

Solve.

9. $5 - 4 =$ _____

10. $9 - 8 =$ _____

11. $6 - 6 =$ _____

12. $10 - 1 =$ _____

13. $7 - 6 =$ _____   14. $6 - 1 =$ _____

15. $3 - 3 =$ _____   16. $0 - 0 =$ _____

Solve for the unknown. Use the blocks if needed.

17. $Y + 7 = 7$ 18. $R + 1 = 3$

Choose the right answer and write it in the blank.

19. Ten minus one equals _____.

zero	six
one	seven
two	eight
three	nine
four	ten
five	

20. Chris found four dimes. He lost zero of them. How many dimes does Chris have now?

_____

Subtract to find the difference. Use addition to check the answer.

1. $\begin{array}{r} 9 \\ -\ 2 \\ \hline \end{array}$

2. $\begin{array}{r} 8 \\ -\ 6 \\ \hline \end{array}$

3. $\begin{array}{r} 3 \\ -\ 2 \\ \hline \end{array}$

4. $\begin{array}{r} 9 \\ -\ 7 \\ \hline \end{array}$

5. $\begin{array}{r} 5 \\ -\ 2 \\ \hline \end{array}$

6. $\begin{array}{r} 6 \\ -\ 4 \\ \hline \end{array}$

7. $\begin{array}{r} 7 \\ -\ 2 \\ \hline \end{array}$

8. $\begin{array}{r} 7 \\ -\ 5 \\ \hline \end{array}$

Solve.

9. $6 - 2 = $ _____

10. $10 - 8 = $ _____

11. $2 - 2 = $ _____

12. $7 - 1 = $ _____

Solve.

13.
$$
\begin{array}{r}
7 \\
+\ 9 \\
\hline
\end{array}
$$

14.
$$
\begin{array}{r}
9 \\
+\ 2 \\
\hline
\end{array}
$$

15.
$$
\begin{array}{r}
4 \\
+\ 9 \\
\hline
\end{array}
$$

16.
$$
\begin{array}{r}
9 \\
+\ 9 \\
\hline
\end{array}
$$

Solve for the unknown. Use the blocks if needed.

17. $B + 5 = 14$

18. $X + 8 = 17$

19. Eight children came to the picnic. Only two of them brought their baseball gloves. How many did not bring baseball gloves?

   _____

20. Emily and her friend played a game. Emily had 7 points, and her friend had 5 points. What is the difference between their scores?

   _____

21

Solve.

1.
$$\begin{array}{r} 11 \\ -\ 9 \\ \hline \end{array}$$

2.
$$\begin{array}{r} 10 \\ -\ 9 \\ \hline \end{array}$$

3.
$$\begin{array}{r} 14 \\ -\ 9 \\ \hline \end{array}$$

4.
$$\begin{array}{r} 17 \\ -\ 9 \\ \hline \end{array}$$

5.
$$\begin{array}{r} 12 \\ -\ 9 \\ \hline \end{array}$$

6.
$$\begin{array}{r} 13 \\ -\ 9 \\ \hline \end{array}$$

7.
$$\begin{array}{r} 18 \\ -\ 9 \\ \hline \end{array}$$

8.
$$\begin{array}{r} 16 \\ -\ 9 \\ \hline \end{array}$$

9. $15 - 9 =$ _____

10. $9 - 9 =$ _____

11. $9 - 2 =$ _____

12. $8 - 6 =$ _____

Solve.

13.
$$5 + 8$$

14.
$$8 + 7$$

15.
$$4 + 8$$

16.
$$800 + 100$$

Solve for the unknown.

17. Q + 6 = 14

18. D + 8 = 11

19. Grace Joy counted 11 cars going by her house. Two of them were red. How many cars were not red?

_____

20. David earned three dollars yesterday doing chores. Today he earned five dollars. How much more money did he earn today than he earned yesterday?

_____

22

Solve.

1.
$$\begin{array}{r} 12 \\ -8 \\ \hline \end{array}$$

2.
$$\begin{array}{r} 15 \\ -8 \\ \hline \end{array}$$

3.
$$\begin{array}{r} 13 \\ -8 \\ \hline \end{array}$$

4.
$$\begin{array}{r} 14 \\ -8 \\ \hline \end{array}$$

5.
$$\begin{array}{r} 16 \\ -8 \\ \hline \end{array}$$

6.
$$\begin{array}{r} 10 \\ -8 \\ \hline \end{array}$$

7.
$$\begin{array}{r} 17 \\ -8 \\ \hline \end{array}$$

8.
$$\begin{array}{r} 9 \\ -8 \\ \hline \end{array}$$

9. $15 - 8 = $ _____

10. $12 - 9 = $ _____

11. $16 - 9 = $ _____

12. $14 - 9 = $ _____

Solve.

13.
$$\begin{array}{r} 4 \\ + \quad 4 \\ \hline \end{array}$$

14.
$$\begin{array}{r} 7 \\ + \quad 7 \\ \hline \end{array}$$

15.
$$\begin{array}{r} 6 \\ + \quad 6 \\ \hline \end{array}$$

16.
$$\begin{array}{r} 300 \\ + \ 300 \\ \hline \end{array}$$

Solve for the unknown.

17. $A + 2 = 9$

18. $X + 1 = 7$

19. Kim has a book with 11 chapters. She has read 8 chapters. How many chapters does she have left to read?

_____

20. Casey needs 17 dollars to buy a gift for her mom. She has 8 dollars. How many more dollars does she need?

_____

23

Solve.

1.
$$\begin{array}{r} 10 \\ -5 \\ \hline \end{array}$$

2.
$$\begin{array}{r} 2 \\ -1 \\ \hline \end{array}$$

3.
$$\begin{array}{r} 14 \\ -7 \\ \hline \end{array}$$

4.
$$\begin{array}{r} 60 \\ -30 \\ \hline \end{array}$$

5.
$$\begin{array}{r} 4 \\ -2 \\ \hline \end{array}$$

6.
$$\begin{array}{r} 16 \\ -8 \\ \hline \end{array}$$

7.
$$\begin{array}{r} 12 \\ -6 \\ \hline \end{array}$$

8.
$$\begin{array}{r} 8 \\ -4 \\ \hline \end{array}$$

9. $18 - 9 = $ _____

10. $17 - 8 = $ _____

11. $12 - 8 = $ _____

12. $15 - 9 = $ _____

Solve.

13.
$$4 + 5$$

14.
$$9 + 1$$

15.
$$3 + 7$$

16.
$$8 + 2$$

Solve for the unknown.

17. A + 5 = 10

18. B + 9 = 16

19. Hannah must work for 6 hours. She has worked for 3 hours. How many more hours does she need to work?

_____

20. It is 12 miles to town. I have 6 miles left to go. How many miles have I already traveled?

_____

24

Solve.

1.
$$10 - 3$$

2.
$$10 - 5$$

3.
$$10 - 2$$

4.
$$10 - 4$$

5.
$$10 - 6$$

6.
$$10 - 8$$

7.
$$10 - 7$$

8.
$$10 - 1$$

9. $16 - 8 =$ _____

10. $6 - 3 =$ _____

11. $14 - 7 =$ _____

12. $18 - 9 =$ _____

Solve.

13.
$$5$$
$$+\ \ 4$$

14.
$$7$$
$$+\ \ 3$$

15.
$$2$$
$$+\ \ 7$$

16.
$$3$$
$$+\ \ 6$$

Solve for the unknown.

17. $A + 1 = 9$

18. $B + 5 = 9$

19. Ethan is 10 years old, and Luke is 4 years old. What is the difference in their ages?

_____

20. Joseph is planning to make ten Christmas cards. Seven of them are finished. How many more cards should he make?

_____

Solve.

1.
$$\begin{array}{r} 13 \\ -9 \\ \hline \end{array}$$

2.
$$\begin{array}{r} 11 \\ -8 \\ \hline \end{array}$$

3.
$$\begin{array}{r} 11 \\ -9 \\ \hline \end{array}$$

4.
$$\begin{array}{r} 14 \\ -9 \\ \hline \end{array}$$

5.
$$\begin{array}{r} 3 \\ -1 \\ \hline \end{array}$$

6.
$$\begin{array}{r} 8 \\ -4 \\ \hline \end{array}$$

7.
$$\begin{array}{r} 7 \\ -1 \\ \hline \end{array}$$

8.
$$\begin{array}{r} 4 \\ -0 \\ \hline \end{array}$$

9.
$$\begin{array}{r} 6 \\ -2 \\ \hline \end{array}$$

10.
$$\begin{array}{r} 9 \\ -9 \\ \hline \end{array}$$

11.
$$\begin{array}{r} 9 \\ -\ 8 \\ \hline \end{array}$$

12.
$$\begin{array}{r} 10 \\ -\ 5 \\ \hline \end{array}$$

13.
$$\begin{array}{r} 3 \\ -\ 2 \\ \hline \end{array}$$

14.
$$\begin{array}{r} 7 \\ -\ 2 \\ \hline \end{array}$$

15.
$$\begin{array}{r} 15 \\ -\ 9 \\ \hline \end{array}$$

16.
$$\begin{array}{r} 2 \\ -\ 1 \\ \hline \end{array}$$

17. $16 - 9 = $ _____

18. $6 - 3 = $ _____

19. $10 - 9 = $ _____

20. $8 - 2 = $ _____

21. $18 - 9 = $ _____

22. $5 - 1 = $ _____

23. $17 - 8 = $ _____

24. $13 - 8 = $ _____

Solve.

25.
$$\begin{array}{r} 7 \\ -5 \\ \hline \end{array}$$

26.
$$\begin{array}{r} 8 \\ -7 \\ \hline \end{array}$$

27.
$$\begin{array}{r} 17 \\ -9 \\ \hline \end{array}$$

28.
$$\begin{array}{r} 14 \\ -8 \\ \hline \end{array}$$

29.
$$\begin{array}{r} 5 \\ -3 \\ \hline \end{array}$$

30.
$$\begin{array}{r} 0 \\ -0 \\ \hline \end{array}$$

31.
$$\begin{array}{r} 9 \\ -2 \\ \hline \end{array}$$

32.
$$\begin{array}{r} 7 \\ -6 \\ \hline \end{array}$$

33.
$$\begin{array}{r} 14 \\ -7 \\ \hline \end{array}$$

34.
$$\begin{array}{r} 12 \\ -9 \\ \hline \end{array}$$

35.
$$\begin{array}{r} 6 \\ -\ 4 \\ \hline \end{array}$$

36.
$$\begin{array}{r} 12 \\ -\ 8 \\ \hline \end{array}$$

37.
$$\begin{array}{r} 6 \\ -\ 5 \\ \hline \end{array}$$

38.
$$\begin{array}{r} 16 \\ -\ 8 \\ \hline \end{array}$$

39.
$$\begin{array}{r} 5 \\ -\ 4 \\ \hline \end{array}$$

40.
$$\begin{array}{r} 2 \\ -\ 2 \\ \hline \end{array}$$

41. $6 - 0 =$ _____

42. $8 - 8 =$ _____

43. $8 - 6 =$ _____

44. $12 - 6 =$ _____

45. $4 - 3 =$ _____

46. $9 - 7 =$ _____

47. $13 - 8 =$ _____

48. $7 - 0 =$ _____

Solve.

1.
$$\begin{array}{r} 9 \\ -\ 4 \\ \hline \end{array}$$

2.
$$\begin{array}{r} 9 \\ -\ 7 \\ \hline \end{array}$$

3.
$$\begin{array}{r} 9 \\ -\ 6 \\ \hline \end{array}$$

4.
$$\begin{array}{r} 9 \\ -\ 0 \\ \hline \end{array}$$

5.
$$\begin{array}{r} 9 \\ -\ 2 \\ \hline \end{array}$$

6.
$$\begin{array}{r} 9 \\ -\ 3 \\ \hline \end{array}$$

7.
$$\begin{array}{r} 9 \\ -\ 1 \\ \hline \end{array}$$

8.
$$\begin{array}{r} 9 \\ -\ 8 \\ \hline \end{array}$$

9. $10 - 6 = $ _____

10. $16 - 8 = $ _____

11. $10 - 7 = $ _____

12. $14 - 7 = $ _____

Solve.

13.
$$4$$
$$+ \quad 3$$

14.
$$3$$
$$+ \quad 5$$

15.
$$8$$
$$+ \quad 5$$

16.
$$9$$
$$+ \quad 4$$

Solve for the unknown.

17. $Q + 1 = 5$

18. $R + 3 = 7$

19. There are 9 people on our team, but 4 of them are sick. How many people are able to play?

_____

20. Mom said that Mike could invite 9 people to his party. He has invited 6 people so far. How many more can he invite?

_____

Solve.

1.
$$\begin{array}{r} 7 \\ -3 \\ \hline \end{array}$$

2.
$$\begin{array}{r} 8 \\ -5 \\ \hline \end{array}$$

3.
$$\begin{array}{r} 7 \\ -4 \\ \hline \end{array}$$

4.
$$\begin{array}{r} 8 \\ -3 \\ \hline \end{array}$$

5.
$$\begin{array}{r} 9 \\ -5 \\ \hline \end{array}$$

6.
$$\begin{array}{r} 9 \\ -6 \\ \hline \end{array}$$

7.
$$\begin{array}{r} 15 \\ -8 \\ \hline \end{array}$$

8.
$$\begin{array}{r} 70 \\ -30 \\ \hline \end{array}$$

9. $12 - 9 = $ _____

10. $17 - 8 = $ _____

11. $14 - 7 = $ _____

12. $9 - 2 = $ _____

Solve.

13.
$$7 \atop{+\ 4}$$

14.
$$6 \atop{+\ 7}$$

15.
$$7 \atop{+\ 5}$$

16.
$$9 \atop{+\ 7}$$

Solve for the unknown.

17. $T + 7 = 15$

18. $P + 3 = 8$

19. We got 7 inches of rain last month and 4 inches this month. How much more rain did we get last month than this month?

_____

20. Anna is 3 years younger than her brother. Her brother is 8. How old is Anna?

_____

Solve.

1.
$$\begin{array}{r} 16 \\ -7 \\ \hline \end{array}$$

2.
$$\begin{array}{r} 13 \\ -7 \\ \hline \end{array}$$

3.
$$\begin{array}{r} 15 \\ -7 \\ \hline \end{array}$$

4.
$$\begin{array}{r} 11 \\ -7 \\ \hline \end{array}$$

5.
$$\begin{array}{r} 12 \\ -7 \\ \hline \end{array}$$

6.
$$\begin{array}{r} 7 \\ -3 \\ \hline \end{array}$$

7.
$$\begin{array}{r} 8 \\ -5 \\ \hline \end{array}$$

8.
$$\begin{array}{r} 70 \\ -40 \\ \hline \end{array}$$

9. $8 - 3 =$ _____

10. $9 - 5 =$ _____

11. $10 - 6 =$ _____

12. $16 - 8 =$ _____

Solve.

13.
$$\begin{array}{r} 5 \\ + \quad 6 \\ \hline \end{array}$$

14.
$$\begin{array}{r} 6 \\ + \quad 9 \\ \hline \end{array}$$

15.
$$\begin{array}{r} 8 \\ + \quad 6 \\ \hline \end{array}$$

16.
$$\begin{array}{r} 6 \\ + \quad 7 \\ \hline \end{array}$$

Solve for the unknown.

17. $Q + 4 = 11$

18. $W + 9 = 18$

19. Kay had 15 dollars in her pocket. She spent 7 dollars. How much money does she have left?

_____

20. Thirteen birds landed on a wire, and then 7 flew away. How many birds are left on the wire?

_____

LESSON TEST

Solve.

1. $\begin{array}{r} 13 \\ -6 \\ \hline \end{array}$

2. $\begin{array}{r} 15 \\ -6 \\ \hline \end{array}$

3. $\begin{array}{r} 11 \\ -6 \\ \hline \end{array}$

4. $\begin{array}{r} 14 \\ -6 \\ \hline \end{array}$

5. $\begin{array}{r} 15 \\ -7 \\ \hline \end{array}$

6. $\begin{array}{r} 8 \\ -3 \\ \hline \end{array}$

7. $\begin{array}{r} 13 \\ -7 \\ \hline \end{array}$

8. $\begin{array}{r} 70 \\ -30 \\ \hline \end{array}$

9. $9 - 7 = $ _____

10. $8 - 1 = $ _____

11. $15 - 9 = $ _____

12. $11 - 8 = $ _____

Solve.

13.
$$\begin{array}{r} 7 \\ +\ 5 \\ \hline \end{array}$$

14.
$$\begin{array}{r} 5 \\ +\ 9 \\ \hline \end{array}$$

15.
$$\begin{array}{r} 8 \\ +\ 5 \\ \hline \end{array}$$

16.
$$\begin{array}{r} 6 \\ +\ 5 \\ \hline \end{array}$$

Solve for the unknown.

17. $A + 6 = 12$

18. $X + 4 = 13$

19. Clara was washing the dishes. She washed 11 plates, but she broke 6 of them. How many plates are left?

_____

20. Ben made 13 wooden toys to sell. He sold 6 of them. How many are left to sell?

_____

Solve.

1.
$$\begin{array}{r} 14 \\ -\ 5 \\ \hline \end{array}$$

2.
$$\begin{array}{r} 12 \\ -\ 5 \\ \hline \end{array}$$

3.
$$\begin{array}{r} 13 \\ -\ 5 \\ \hline \end{array}$$

4.
$$\begin{array}{r} 11 \\ -\ 5 \\ \hline \end{array}$$

5.
$$\begin{array}{r} 13 \\ -\ 6 \\ \hline \end{array}$$

6.
$$\begin{array}{r} 12 \\ -\ 7 \\ \hline \end{array}$$

7.
$$\begin{array}{r} 11 \\ -\ 6 \\ \hline \end{array}$$

8.
$$\begin{array}{r} 40 \\ -\ 20 \\ \hline \end{array}$$

9. $16 - 7 =$ _____

10. $14 - 6 =$ _____

11. $13 - 7 =$ _____

12. $15 - 6 =$ _____

Solve.

13.   7
    +   4
    _____

14.   4
    +   9
    _____

15.   9
    +   3
    _____

16.   8
    +   3
    _____

Solve for the unknown.

17. $D + 6 = 11$ 18. $F + 8 = 12$

19. Katie needs to learn 12 new words for science class. She has learned 5 of the words. How many are left to learn?

    _____

20. The red picture book has 14 pages, and the green book has 5 pages. How many more pages does the red book have?

    _____

Solve.

1.
$$\begin{array}{r} 1\,2 \\ -4 \\ \hline \end{array}$$

2.
$$\begin{array}{r} 1\,1 \\ -3 \\ \hline \end{array}$$

3.
$$\begin{array}{r} 1\,3 \\ -4 \\ \hline \end{array}$$

4.
$$\begin{array}{r} 1\,2 \\ -3 \\ \hline \end{array}$$

5.
$$\begin{array}{r} 1\,1 \\ -4 \\ \hline \end{array}$$

6.
$$\begin{array}{r} 1\,2 \\ -5 \\ \hline \end{array}$$

7.
$$\begin{array}{r} 1\,1 \\ -7 \\ \hline \end{array}$$

8.
$$\begin{array}{r} 80 \\ -\,50 \\ \hline \end{array}$$

9. $11 - 5 =$ _____

10. $15 - 7 =$ _____

11. $14 - 6 =$ _____

12. $12 - 7 =$ _____

13. 13 − 5 = _____   14. 15 − 6 = _____

Solve.

15.   8   16.   6
        + 4   + 6

17.   7   18.   9
        + 7   + 2

19. Elizabeth solved 13 math problems. She checked her
     answers, and 5 were wrong. How many answers were
     correct?

     _____

20. Maddie baked 11 cookies. Three of them are burned. How
     many cookies are not burned?

     _____

IV

Solve.

1. $9 - 4 = $ _____

2. $7 - 4 = $ _____

3. $11 - 3 = $ _____

4. $11 - 6 = $ _____

5. $13 - 7 = $ _____

6. $12 - 7 = $ _____

7.
$$\begin{array}{r} 13 \\ -5 \\ \hline \end{array}$$

8.
$$\begin{array}{r} 11 \\ -4 \\ \hline \end{array}$$

9.
$$\begin{array}{r} 9 \\ -6 \\ \hline \end{array}$$

10.
$$\begin{array}{r} 7 \\ -3 \\ \hline \end{array}$$

11.
$$\begin{array}{r} 11 \\ -5 \\ \hline \end{array}$$

12.
$$\begin{array}{r} 13 \\ -6 \\ \hline \end{array}$$

13.
$$\begin{array}{r} 12 \\ -4 \\ \hline \end{array}$$

14.
$$\begin{array}{r} 8 \\ -3 \\ \hline \end{array}$$

15.
$$\begin{array}{r} 8 \\ -5 \\ \hline \end{array}$$

16.
$$\begin{array}{r} 11 \\ -7 \\ \hline \end{array}$$

17.
$$\begin{array}{r} 12 \\ -3 \\ \hline \end{array}$$

18.
$$\begin{array}{r} 15 \\ -7 \\ \hline \end{array}$$

19.
$$\begin{array}{r} 14 \\ -5 \\ \hline \end{array}$$

20.
$$\begin{array}{r} 9 \\ -3 \\ \hline \end{array}$$

21. $15 - 6 =$ _____

22. $16 - 7 =$ _____

23. $9 - 5 =$ _____

24. $14 - 6 =$ _____

25. $12 - 5 =$ _____

26. $13 - 4 =$ _____

FINAL TEST

Solve.

1.
$$\begin{array}{r} 10 \\ -3 \\ \hline \end{array}$$

2.
$$\begin{array}{r} 7 \\ +3 \\ \hline \end{array}$$

3.
$$\begin{array}{r} 8 \\ -4 \\ \hline \end{array}$$

4.
$$\begin{array}{r} 4 \\ +7 \\ \hline \end{array}$$

5.
$$\begin{array}{r} 9 \\ -6 \\ \hline \end{array}$$

6.
$$\begin{array}{r} 9 \\ +9 \\ \hline \end{array}$$

7.
$$\begin{array}{r} 12 \\ -7 \\ \hline \end{array}$$

8.
$$\begin{array}{r} 8 \\ +7 \\ \hline \end{array}$$

9. $\begin{array}{r} 15 \\ -9 \\ \hline \end{array}$

10. $\begin{array}{r} 12 \\ -4 \\ \hline \end{array}$

11. $\begin{array}{r} 5 \\ +3 \\ \hline \end{array}$

12. $\begin{array}{r} 13 \\ -6 \\ \hline \end{array}$

13. $\begin{array}{r} 10 \\ -5 \\ \hline \end{array}$

14. $\begin{array}{r} 7 \\ +6 \\ \hline \end{array}$

15. $\begin{array}{r} 3 \\ +6 \\ \hline \end{array}$

16. $\begin{array}{r} 11 \\ -8 \\ \hline \end{array}$

17.
$$
\begin{array}{r}
8 \\
+\ 5 \\
\hline
\end{array}
$$

18.
$$
\begin{array}{r}
4 \\
+\ 9 \\
\hline
\end{array}
$$

19.
$$
\begin{array}{r}
17 \\
-\ 9 \\
\hline
\end{array}
$$

20.
$$
\begin{array}{r}
14 \\
-\ 5 \\
\hline
\end{array}
$$

21.
$$
\begin{array}{r}
3 \\
+\ 8 \\
\hline
\end{array}
$$

22.
$$
\begin{array}{r}
13 \\
-\ 9 \\
\hline
\end{array}
$$

23.
$$
\begin{array}{r}
5 \\
+\ 7 \\
\hline
\end{array}
$$

24.
$$
\begin{array}{r}
16 \\
-\ 7 \\
\hline
\end{array}
$$

25.
$$\begin{array}{r} 9 \\ + \quad 3 \\ \hline \end{array}$$

26.
$$\begin{array}{r} 11 \\ - \quad 6 \\ \hline \end{array}$$

27.
$$\begin{array}{r} 15 \\ - \quad 8 \\ \hline \end{array}$$

28.
$$\begin{array}{r} 7 \\ + \quad 4 \\ \hline \end{array}$$

29.
$$\begin{array}{r} 5 \\ + \quad 6 \\ \hline \end{array}$$

30.
$$\begin{array}{r} 8 \\ + \quad 7 \\ \hline \end{array}$$